The LEAN Playbook Series

The Kanban Playbook

A Step-by-Step Guideline for the Lean Practitioner

The LEAN Playbook Series

PUBLISHED

The 5S Playbook: A Step-by-Step Guideline for the Lean Practitioner
Chris A. Ortiz

The Kanban Playbook: A Step-by-Step Guideline for the Lean Practitioner
Chris A. Ortiz

FORTHCOMING

The Quick Changeover Playbook: A Step-by-Step Guideline for the Lean Practitioner
Chris A. Ortiz

The TPM Playbook: A Step-by-Step Guideline for the Lean Practitioner
Chris A. Ortiz

The Cell Manufacturing Playbook: A Step-by-Step Guideline for the Lean Practitioner
Chris A. Ortiz

The LEAN Playbook Series

The Kanban Playbook

A Step-by-Step Guideline for the Lean Practitioner

Chris A. Ortiz

CRC Press
Taylor & Francis Group
Boca Raton   London   New York

CRC Press is an imprint of the
Taylor & Francis Group, an **informa** business

A PRODUCTIVITY PRESS BOOK

CRC Press
Taylor & Francis Group
6000 Broken Sound Parkway NW, Suite 300
Boca Raton, FL 33487-2742

© 2016 by Taylor & Francis Group, LLC
CRC Press is an imprint of Taylor & Francis Group, an Informa business

No claim to original U.S. Government works

Printed on acid-free paper
Version Date: 20150714

International Standard Book Number-13: 978-1-4987-4175-0 (Paperback)

This book contains information obtained from authentic and highly regarded sources. Reasonable efforts have been made to publish reliable data and information, but the author and publisher cannot assume responsibility for the validity of all materials or the consequences of their use. The authors and publishers have attempted to trace the copyright holders of all material reproduced in this publication and apologize to copyright holders if permission to publish in this form has not been obtained. If any copyright material has not been acknowledged please write and let us know so we may rectify in any future reprint.

Except as permitted under U.S. Copyright Law, no part of this book may be reprinted, reproduced, transmitted, or utilized in any form by any electronic, mechanical, or other means, now known or hereafter invented, including photocopying, microfilming, and recording, or in any information storage or retrieval system, without written permission from the publishers.

For permission to photocopy or use material electronically from this work, please access www.copyright.com (http://www.copyright.com/) or contact the Copyright Clearance Center, Inc. (CCC), 222 Rosewood Drive, Danvers, MA 01923, 978-750-8400. CCC is a not-for-profit organization that provides licenses and registration for a variety of users. For organizations that have been granted a photocopy license by the CCC, a separate system of payment has been arranged.

Trademark Notice: Product or corporate names may be trademarks or registered trademarks, and are used only for identification and explanation without intent to infringe.

Library of Congress Cataloging-in-Publication Data

Ortiz, Chris A.
  The Kanban playbook : a step-by-step guideline for the lean practitioner / Chris A. Ortiz.
    pages cm
  Includes index.
  ISBN 978-1-4987-4175-0
  1. Production control. 2. Just-in-time systems. 3. Production management. I. Title.

TS157.O78 2016
658.5--dc23   2015018290

Visit the Taylor & Francis Web site at
http://www.taylorandfrancis.com

and the CRC Press Web site at
http://www.crcpress.com

Contents

How to Use This Playbook vii
Introduction ix

1 Cycle Counting and Baseline 1
    Introduction 1
    Creating a Kanban Sizing Report 2
    Cycle Counting, Part Identification, Vendors, and On-Hand Inventory 3
    Identification of New Inventory Quantities 4
    Calculating New On-Hand Costs 5
        Creating a Kanban Sizing Report 6

2 5S and New Maximum Quantities 7
    Sorting 8
    Location Designations 13

3 Kanban Cards 15
    Introduction 15
    Benefits of Kanban 15
    Kanban Defined 16
    Creating a Kanban Card System 17
    Break Down Items into Categories 17
    Establish Maximum Quantity: Usage Is Key 18
    Pros and Cons of Higher Maximum Quantities 18
    Pros and Cons of Lower Maximum Quantities 18
    Maximum Quantity Examples 18
    Establish the Minimum Quantity 19
    Establish Reorder Signal or Quantity 19
    Reorder Signal 20
    Kanban System for Supplies 21
    Supplies Are Delivered and Card Returned 23
    Maintenance Kanban Cards 25

4 Two-Bin Systems and Material Handling 27
    Two-Bin Systems 27
    Material Handling 32
        Communication Lights 33
        Kanban Card or Bin Drop Point 34

5 In-Process Kanban 37
    What Is a Pull System? 37
        External and Internal Customers 37
        Process Based (Push) 38
        In-Process Kanban (Pull) 39
    Pull System from Assembly to Paint 40
    Items Have Location on Rack 40
    Information and Signal 41
    How It Works 42
    Welding into Powder Coating 46
    IPK into Powder Coating Department 47
    Writable In-Process Kanban (IPK) 48
        Saw Operation 48

Conclusion 49

Definition of Terms 51

Index 53

About the Author 59

How to Use This Playbook

In most cases, a playbook is a spiral-bound notebook that outlines a strategy for a sport or a game. Whether it is a football game, a video game, or even a board game, playbooks are all around us, and when written properly provide immediate and easily understood direction. Playbooks can also provide general information; then, it is up to the users of the playbook to tailor it to their individual needs.

Playbooks contain pictures, diagrams, quick reference, definitions, and often step-by-step illustrations to explain certain parts. You can use playbooks to either help you understand the entire game or you can pick and choose and focus on one element. The bottom line is that any playbook should be easy to read and to the point and contain little to no filler information.

The Kanban Playbook is written for the Lean practitioner and facilitator. Like a football coach, a facilitator can use this playbook for quick reference and then be able to convey what is needed easily. If for some reason the person leading the actual Kanban implementation forgets a "play," the person can reference the playbook.

You can either follow page by page and use it to facilitate a Kanban implementation or you can go directly to certain topics and use it to help you implement that particular play.

How to Use This Playbook

Introduction

At first glance, the improvement techniques within the Lean philosophy appear to provide a solution to many types of production-related issues. A powerful and effective improvement philosophy, Lean can prevent company failure or launch an organization into world-class operational excellence.

I have been a Lean practitioner for over 15 years and have been involved in many Lean transformations. It does not matter the industry you work in, the product you produce, or even the processes your company uses to transform a finished good; the problems and opportunities you face are the same as everyone else. Your company is not "different" or the exception to everyone else. You, as a Lean practitioner, desire a smoother-running facility, reduced lead times, more capacity, improved productivity, flexible processes, usable floor space, reduced inventory, and so on. Organizations can implement Lean to make localized improvements or they can make Lean transform the entire culture of the business. Regardless of your aspirations and goals for Lean, you and many other companies face another similar situation: getting out of what I call *boardroom Lean* and moving toward implementation.

Have no illusions: Lean is about rolling your sleeves up, getting dirty, and making change. True change comes on the production floor, in the maintenance shop, and in all the other areas of the organization and implementing the concepts of Lean. Companies often get stuck in endless cycles of training and planning, with no implementation ever happening. This playbook is your guideline for implementation and is written for the pure Lean practitioner looking for a training tool and a guideline that can be used in the work area while improvements are being conducted. There is no book, manual, or reference guide that provides color images and detailed step-by-step guidelines on how to properly implement Kanban. The implementation of Kanban is a manually intensive action, and conducting Kanban projects properly takes experience and direction. *The Kanban Playbook* is not a traditional book, as you can probably see. It is not intended to be read like another Lean business book. The images in this playbook are from real Kanban implementations, and I use a combination of short paragraphs and bulleted descriptions to walk you through how to effectively implement Kanban.

Little or no time is wasted on high-level theory, although an introductory portion is dedicated to the 8 Wastes and Lean metrics is included. An understanding of wastes and metrics is needed to fully benefit from this playbook. Not implying high-level theory or business strategies are not valuable—they are highly valuable. This playbook is for implementation, so it will not contain filler information.

When discussing the concept of Kanban, we are talking about inventory and creating a signal system to trigger the need for it. The term *inventory* is broad, but for the sake of this playbook, the focus is on the system needed between the warehouse or stockroom and the movement of WIP (work in process) between work areas. Ideally, cell manufacturing should be incorporated whenever possible and there is another playbook in this series that discusses it. However, in some cases cell manufacturing may not be feasible. This is not always the case, so this playbook can be used if you will not be using cell manufacturing and need some insight on how to reduce WIP and move it through the plant. This playbook does not spend a lot of time on the external supply chain. This is not to imply that developing Kanban for the external side of the supply chain is not valuable; it is more time intensive, and it is best to use traditional business books that have been written to explain the concept.

Chapter 1 contains the fewest pictures of all the chapters; a series of diagrams and forms is used to illustrate how to cycle count and baseline inventory in a work area. We start in the work area as that is where the value-added work is performed, and it is best to start from that point and move back pulling material, parts, and supplies into the area. This chapter provides information on creating a Kanban sizing report, cycle counting, part identification, minimum and maximum quantities, and calculating current and future on-hand inventory cost.

Chapter 2 sets the tone for a successful Kanban implementation by showing you how to use 5S (a concept of a visual organization that creates home locations for all essential items) and the visual workplace in establishing how much space will be needed for the Kanban system. By using the established inventory levels created in Chapter 1, you will see how using 5S in connection with Kanban can make a more organized and effective system.

The next chapter focuses on the creation and placement of Kanban cards. These cards are the reorder mechanisms in the Kanban system; Chapter 3 illustrates how to design and create the cards/lists. Real-life examples and multiple options are shown to help point you in a direction that makes sense for your process. Examples are from production, maintenance, and shipping areas. Chapter 4 is a unique chapter as it explains the concept of a two-bin system and material handling. Two-bin systems are Kanban processes that do not utilize cards for reordering, but the bins themselves that hold and store parts and material. It is a great alternative in production areas because it can help monitor output and the flow of the work. A step-by-step illustration on how the system works when in use is provided.

With a firm understanding of how to implement Kanban, Chapter 4 also discusses material handling and the process in which people are involved in using the Kanban system. Material-handling processes are an essential part of any Lean process. This chapter discusses the role of the material handler and provides examples of how to work in this reordering system, milk runs, rotations, drop points, and replenishment. Various diagrams are used to help illustrate the concept.

Chapter 5 moves away from raw material and parts and focuses on WIP Kanbans or what is called In-Process Kanbans (IPKs). Kanban is applicable when dealing with WIP. When cell manufacturing is not an option, IPKs help you convert traditional push systems into pull systems. By implementing an IPK system, a Lean practitioner can reduce inventory, develop pull, create visibility, and decrease lead times throughout the flow of the process. Multiple examples are provided.

Kanban systems are effective when planned and implemented properly. This playbook provides the most visible and detailed approach to Kanban implementation so you, the Lean practitioner, can see results in a short period. Like playbooks in this series, use it on the work area and lead your Kanban teams to success.

8 Wastes

As a Lean practitioner and teacher myself, I know the power of Kanban, and when you challenge your viewpoint and the handling of inventory, you can see remarkable changes in your overall company. Like any Lean tool, Kanban can have a significant impact on waste, and it is good to refresh your memory on the 8 Wastes. Many of you reading this playbook already understand the concepts of waste and Lean. For those of you just getting started, the following are the 8 Wastes.

Overproduction
Overprocessing
Waiting
Motion
Transportation
Inventory
Defects
Wasted potential

Overproduction is the act of making more product than necessary and completing it faster than necessary and before it is needed. Overproduced product takes up floor space, requires handling and storage, and could result in potential quality problems if the lot contains defects.

Overprocessing is the practice of extra steps, rechecking, reverifying, and outperforming work. Overprocessing is often conducted in fabrication departments when sanding, deburring, cleaning, or polishing is overperformed. Machines can also over-process when they are not properly maintained and simply take more time to produce quality parts.

Waiting occurs when important information, tools, and supplies are not readily available, causing machines and people to be idle. Imbalances in workloads and cycle times between processes can also cause waiting.

Motion is the movement of people in and around the work area to look for tools, parts, information, people, and all necessary items that are not available. When a process contains a high level of motion, lead time increases and the focus on quality begins to decrease. All necessary items should be organized and be placed at point of use so the worker can focus on the work at hand.

Transportation is the movement of parts and product throughout the facility. Often requiring a forklift, hand truck, or pallet jack, transportation exists when consuming processes are far away from each other and are not visible.

Inventory is a waste when manufacturers tie up too much money holding excessive levels of raw, WIP, and finished goods inventory.

Defects are any quality metric that causes rework, scrap, warranty claims, and rework hours from mistakes made in the factory.

Wasted human potential is the act of not properly utilizing employees to the best of their abilities. People are only as successful as the process they are given to work in. If a process inherently has motion, transportation, overprocessing, overproduction, periods of waiting, and defect creation, then that is exactly what they will do. That is wasted human potential.

A Kanban system can help you reduce these eight wastes and by doing so will create a much more productive and profitable company for all.

My hope is that you will read this playbook and not only be inspired, but also be able to roll up your sleeves and begin your Kanban journey after the last page is read.

Lean Metrics

To effectively measure your success with Kanban, you need to establish a list of critical shop-floor metrics that can be measured and quantified. On the production floor, these metrics are often called key performance indicators (KPIs). Kanban is a powerful improvement tool that can have a profound impact on reducing lead times, reducing inventory, increasing output, and improving productivity and many other types of KPIs. In some cases, the change is dramatic. We recommend the following Lean metrics become part of measuring your overall LEAN journey:

- Productivity
- Quality
- Inventory
- Floor space
- Travel distance
- Throughput time

Productivity

Productivity is measured in a variety of different ways. Productivity is improved when products are manufactured with less effort. This reduction in effort essentially is the reduction of waste. Kanban systems can reduce or eliminate all of the steps and time associated with searching for parts and material in an unorganized work area. Once you have reduced or eliminated the time spent searching for supplies and parts and walking long distances, there is more time in the day that can be allocated and focused on performing work. And in this case, it is the value-added work of making products that should be maximized for higher productivity.

Now, the same number of people can produce more work in the same amount of time.

$$\text{Fewer steps} + \text{Same people} = \text{Higher productivity}$$

Quality

Improvements to quality are more of a secondary benefit of Kanban. Kanban implementations have an impact on internal quality, such as rework, defects, scrap, and labor, simply ensuring the right parts are placed in the work area. Kanban systems challenge inventory levels, because less inventory is in the plant, the result is less damage and reduced possibilities of material becoming obsolete due to long-term storage.

Inventory

Inventory is tricky because most companies producing or repairing a product need all three levels of it: raw, WIP, and finished goods. Traditional accounting practices look at inventory as an asset and spend a considerable amount of time trying to reduce the cost per unit. Often, buyers will purchase large quantities of inventory to receive a percentage discount per part. The thinking here is that the cost per unit is going down, and when that part is finally used or sold, there is more profit. There is a problem with that model: carrying costs.

Carrying costs are all the expenses associated with having inventory; this cost is hidden and is not seen on a product costing model. At best, a company may see a 5% discount from bulk purchasing. Once that order arrives, carrying costs kick in, and they do not end until to the inventory is used or sold.

Moving inventory around the company requires people and machines. Inventory takes up floor space, which costs money, and may be taking up space that should be used for something more useful. The company needs to purchase racks and shelving, install systems to monitor the inventory, and have manpower to continually handle it. As inventory levels go up, there is lost labor in moving it around to get to the required inventory. Add some insurance, the occasional damage to material due to excessive handling, and throw in some obsolescence, and you have just taken a 5% discount and created a healthy amount of cost. Inventory is waste, and this playbook provides ways to attack it.

Inventory ties up money, contributes to clutter, takes up floor space, and often provides some of the most common physical obstacles in the company. Workers spend time shifting material and inventory around just to locate what they need. Time is lost by dealing with excessive inventory just to reach the items required to perform their work. As you are organizing the work area, you should be considering how you may ultimately start reducing inventory levels to help create a more visual workplace.

Floor Space

Floor space comes at a premium, and you need to start looking at the poor use of floor space as hurting the company's ability to grow. Floor space should be used to perform value-added work that creates revenue for the company. It should not be used to store junk or act as a collector of unneeded items. Renting, leasing, or buying a manufacturing building is one of the highest costs of overhead. The production floor is in place to serve one purpose: to build products. Although the factory is used for other items, such as holding inventory, shipping, receiving, maintenance, and so on, the production floor should be effectively utilized for value-added work. Value-added work is the act of building products or the steps needed to change fit, form, or function of the product you intend to sell. Production lines, equipment, and machines all produce a sellable product, and the floor space needed to perform this work should be properly used.

Buildings cost money, and there is a lot associated with having a facility even if you produce product but not every month. The costs can include the lease, insurance, taxes, utilities, maintenance, and upkeep. You better be making money out of it. How much of your space is used to create revenue? Inventory sitting on shelves does not create revenue in a production or repair environment. You can measure your floor space utilization by something called revenue or profit per square foot. As you implement Kanban systems, you will see a reduction in overall carrying costs in your facility. One of the costs is poor use of space, and it is a critical Lean metric.

As a company becomes less organized and unneeded inventory begins to accumulate, more space becomes used for non-value-added items. This creates an increase in waste. Over time, other items such as workbenches, garbage cans, chairs, unused equipment, tools, and tables tend to pile up, and valuable production space simply disappears. Rather than reduce inventory and improve floor space use, the general approach is to add. Add building space, racks, and shelves, and you want to change your perception of space to better use, fewer non-value-added items, less waste, and less stuff.

Travel Distance

Here is the best way to view travel distance: the farther it has to go, the longer it is going to take. Long production processes can create a lot of waste and can reduce overall performance. Plus, longer-than-needed processes take up floor space. There are two ways to look at travel distance: the distance people walk and the distance inventory (product) is transported.

Travel distance is connected to overall lead times in a process and the entire factory. When WIP is created above required quantities, it takes up valuable floor space and increases the distance the production line needs. As travel distance increases, floor space becomes improperly used, workers walk further distances, and lead times are increased. Wait time between processes also increases, and there is added lead time to maneuver inventory.

When work areas are designed incorrectly, they can create a lot of walking for workers, and as the area becomes cluttered, more time is needed to find essential items for work.

As waste is reduced through 5S and Kanban, the travel distance of product and workers decreases, making travel distance a good Lean metric.

Throughput Time

Sometimes used in conjunction with measuring travel distance reduction, throughput time is the time it takes the product to flow through the production process. Throughput time has a direct impact on delivery; the longer it takes product to move through the plant, the longer it takes to be delivered. Of course, there are a multitude of variables that can extend product lead time; therefore, it is wise to simplify the metric by measuring the time when process 1 grabs raw material to the time it is packaged and ready for shipment. Longer production lines require more workstations, workers, tools, conveyors, supplies, and material, which results in additional cost and WIP as well as extended lead times. A physical reduction in distance equates to less throughput time, allowing an organization to promise more competitive, yet reasonable, delivery dates. Kanban systems help reduce overall lead times at the front end of the process and within it. Moving inventory faster through the process means less time and money.

Improving these key Lean metrics and using them as a measurement of your success will have a profound impact on the overall financial success and long-term growth of the company. One could look at these Lean metrics simply as process metrics because they can be measured down at the shop-floor level. Production workers need to work in an efficient environment to be successful contributors to optimal cost, quality, and delivery. Each Lean metric improved complements another, and another, and so on. As you become a better Lean practitioner, your understanding of how the metrics relate to each other will become second nature.

Chapter 1

Cycle Counting and Baseline

Introduction

Before you can run out and start implementing a Kanban system, there is some preliminary work that needs to be done. This chapter covers a variety of different planning items that will help implement a more effective and user-friendly Kanban system.

Items that are discussed:

- Creating a Kanban sizing report
- Cycle counting, part identification, vendors, and on-hand inventory
- Identification of inventory quantities
- Calculating on-hand cost

Creating a Kanban Sizing Report

Maintenance Shop Supplies Report

Category	Part Description	Part #	Vendor	Cost per Unit	On Hand	Total on Hand
Fittings	#6 Quick Connects	FF-371-6FP	ACH	$7.50	13	$97.50
Fittings	#6 Quick Connects	FF-372-6FP	ACH	$7.50	13	$97.50
Fittings	Quick Disconnects	STUCCI-M-A7	ACH	$7.50	5	$37.50
Fittings	Quick Disconnects	STUCCI-F-A7	ACH	$7.50	4	$30.00
Fittings	Quick Disconnects	STUCCI-M-FIR614	ACH	$7.50	2	$15.00

A Kanban sizing report is a document that can be created in Microsoft Excel that is used to perform cycle counting and obtain baseline information for the implementation.

Category: Break down your inventory into categories based on the department
Part Description: Description of the supply part that is identified by the vendor
Part #: Actual part number used to order from the vendor
Vendor: Name of supplier you order the part from
Cost per Unit: What is the average cost of the part from the supplier?
On Hand: How many parts are on hand at the moment of cycle counting?
Total Cost on Hand: Multiply the cost per unit times the on-hand quantity

Cycle Counting, Part Identification, Vendors, and On-Hand Inventory

- Staff using Kanban sizing report
- Conducting cycle counts on maintenance supplies and parts

- Tagging parts in the storage area with information on the part
- Helps match up Kanban card with items during implementation

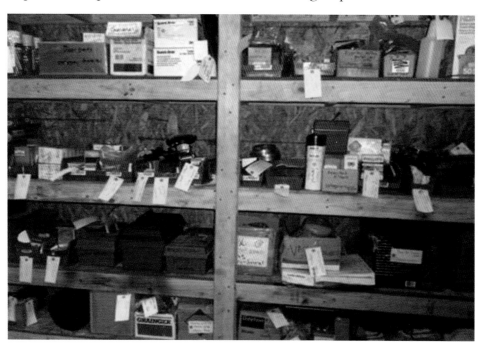

Identification of New Inventory Quantities

Maintenance Shop Supplies Report

Category	Part Description	Part #	Vendor	Cost per Unit	On Hand	Total on Hand	Max Qty	Min Qty	Reorder Qty
Fittings	#6 Quick Connects	FF-371-6FP	ACH	$7.50	13	$97.50	4	1	3
Fittings	#6 Quick Connects	FF-372-6FP	ACH	$7.50	13	$97.50	4	1	3
Fittings	Quick Disconnects	STUCCI-M-A7	ACH	$7.50	5	$37.50	4	1	3
Fittings	Quick Disconnects	STUCCI-F-A7	ACH	$7.50	4	$30.00	3	1	2
Fittings	Quick Disconnects	STUCCI-M-FIR614	ACH	$7.50	2	$15.00	1	0	1

Some people think you need to look at a lot of data to establish new inventory levels. Depending on the situation, you might when it comes to production lines. This is an example from a maintenance department with very sporadic usage information. Simply use the experienced people in the department to establish new inventory levels.

You will have to go through each part and discuss its usage, vendor, and lead time to come up with new maximum and minimum quantities. As odd as it may appear, this process is effective in the first pass, and you can refine your numbers after the implementation.

Max Qty: This establishes the highest level of inventory you want based on usage.

Min Qty: This is the quantity of safety stock to work from while the part is being ordered.

Reorder Qty: Once the minimum quantity is reached, this will trigger the need to turn in the Kanban card; this is the quantity that is ordered to refill to the required maximum.

Calculating New On-Hand Costs

Part #	Vendor	Cost per Unit	On Hand	Total on Hand	Max Qty	Min Qty	Reorder Qty	New Max Cost	Difference
FF-371-6FP	ACH	$7.50	13	$97.50	4	1	3	$30.00	$67.50
FF-372-6FP	ACH	$7.50	13	$97.50	4	1	3	$30.00	$67.50
STUCCI-M-A7	ACH	$7.50	5	$37.50	4	1	3	$30.00	$7.50
STUCCI-F-A7	ACH	$7.50	4	$30.00	3	1	2	$22.50	$7.50
STUCCI-M-FIR614	ACH	$7.50	2	$15.00	1	0	1	$7.50	$7.50
	Total On Hand Before			$277.50	Total On Hand Savings			$120.00	$157.50

The final Kanban sizing report shows the previous on-hand costs of inventory at $277.50 and the new on hand after the implementation will be $120.00. Although this is a simple example, if you follow this thought pattern when designing your Kanban system, it could reach a savings of hundreds of thousands to millions of dollars, depending on the level of inventory and cost per unit.

Creating a Kanban Sizing Report

There are much more complex Kanban sizing reports, and the number of parts will dictate how big the report will be. Now, this completed report is used to help you organize the storage and work area. Use the newly established maximum quantities to dictate the size of bins, shelf and work space, and overall floor space. Once that is complete, you can use the report to create a Kanban system.

Often, the information that is collected during the cycle counting can be obtained from some form of material information system such as MRP, ERP, or some type of maintenance-tracking software. If you have faith in the system's integrity, then go ahead. However, nothing provides better deep insight on the current state, buying habits, size of parts, and unused inventory than sheet cycle counting.

Once you have completed your cycle counting, baseline, and the Kanban sizing report, you are ready to design the Kanban system for implementation.

Chapter 2
5S and New Maximum Quantities

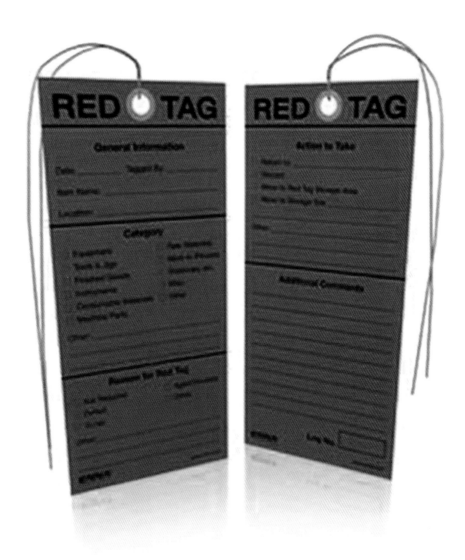

5s is comprised of Sort, Set in Order, Scrub, Standardize, and Sustain.

Sorting

Sort is the act of removing and discarding all unnecessary items from a work area; during a Kanban implementation, your implementation team is sorting unneeded, broken, and obsolete inventory. During this sorting phase, the team is also removing excess quantities above your newly established maximum quantities in the Kanban sizing report. When you reset the inventory quantities per the Kanban sizing report, you simply remove the excess and stage as a "use-up" pile. Employees use these items first; when the inventory is gone, you start the new Kanban system. You can also use red tags to mark the items so there is an organized inventory list for everyone to use during the use-up phase.

- Excessive maintenance supplies above new maximum quantities
- Staged during the right-sizing phase

5S and New Maximum Quantities ■ 9

■ Resetting the storage cabinet

■ Place excess inventory into specific piles

- Extra parts from sorting
- Overproduced fabricated parts

- Excessive corrugated material
- Four-month supply
- Never used and now obsolete

- Extra welded parts
- Some obsolete subassemblies

When establishing new storage needs, always use the maximum quantity identified on the Kanban sizing report. This quantity dictates everything: size of shelf, rack, bins, and the actual space the item takes up.

- Body shop supplies stored back in the parts room
- Minimal space in depth, height, and length of shelf
- No doors: visual and easily retrievable

- Hardware stored in work area
- Minimal space used
- Adjustable shelves to minimize height

- Maintenance parts on a shadow board
- Eliminated cabinet
- Reduced inventory quantity
- Opened up floor space
- Inventory stored on shadow boards

Location Designations

The Kanban sizing report can include location information for each piece of inventory. This information will be printed on the Kanban card as well as on the storage device where it is placed. This provides a quick reference from a distance when replenishing parts and supplies in the work area. The information on the card is the same as the location designation.

- Packaging station
- Four racks dedicated to boxes
- S1, S2, S3, S4: Rack identification. Acts as an "address" for items stored in the rack

Chapter 3

Kanban Cards

Introduction

Kanban cards are a visual signal system that is implemented to trigger the need for inventory. These cards are placed all over the facility as needed and can have a significant impact on output and productivity. Cards can be placed in work areas and used when material and supplies are running low, with the card turned into a material handler to retrieve the needed item. The card contains all the vital information needed, including part number, part description, quantity, and locations.

Kanban card systems can also be used when ordering from outside suppliers but generally take longer to implement. This chapter describes the internal Kanban card system of a company.

Benefits of Kanban

- Reduced inventory
- Reduced cost
- Reduced floor space use
- Reduced motion
- Creates better visibility of shortages
- Helps transform to a pull system
- Controls inventory
- Creates a robust and repeatable system

Foam Tape

Min: 1 Roll
Max: 2 Rolls
Reorder Qty: 1 Roll

WSD

Kanban Defined

- Japanese word for "signal"
- Used to pull and trigger work
    - Moving product through the shop
    - Ordering material and parts
    - Ordering supplies
    - Requesting paperwork
    - Calling for help or assistance

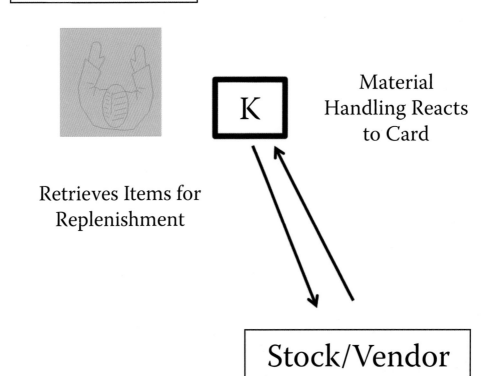

Creating a Kanban Card System

1. Break down items into categories
2. Remove all unused parts, material, supplies
3. Establish maximum quantity: usage is key
4. Establish the minimum quantity
5. Establish the reorder signal or quantity
6. Right size the area (smarter storage size)
7. Implement 5S for organization
8. Make Kanban cards
9. Implement Kanban card system

Break Down Items into Categories

- Parts
  - Brackets
  - Panels
- Material
  - Sheets of metal
- Shop supplies
  - Rags, fillers, adhesives, cleaners, lubricants, tape
- Hardware
  - Nuts/bolts/washers, fasteners, and so on

Establish Maximum Quantity: Usage Is Key

- Get an idea of usage
- Establish the maximum amount on hand
- Use a timeline: 2 days, 3 days, and so on
- Maximum level used to right size the storage

Pros and Cons of Higher Maximum Quantities

- Pro: Comfort level of reduced shortages
- Pro: Lower frequency of material handling
- Con: More inventory
- Con: Lower inventory turns
- Con: Higher cost
- Con: More floor space use in production
- Con: Less visibility of part problems

Pros and Cons of Lower Maximum Quantities

- Pro: Lower inventory
- Pro: Faster inventory turns
- Pro: Less floor space in production
- Pro: Better visibility of part problems
- Con: Less buffer for part shortages

Maximum Quantity Examples

- Bottle quantities

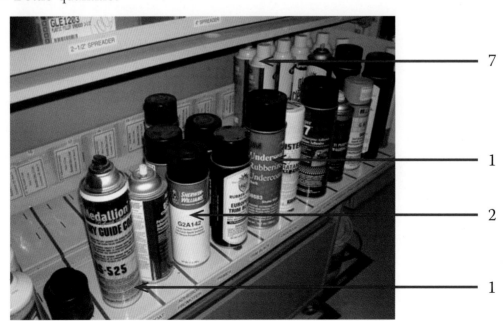

Establish the Minimum Quantity

- Smallest amount on hand
- Acts as a buffer while replenishment takes place

Establish Reorder Signal or Quantity

- Trigger for replenishment
- Signal to turn in card for ordering
- Creates a robust repeatable ordering process
- Signal everyone follows

Reorder Signal

- Cabinet is right size and organized and has Kanban cards.

- Red dot is signal to turn in card.
- Card is turned in when bottle is empty and recycled.

Kanban System for Supplies

- Reassembly area

- Reorder signal. Place card in "Order" box.

- Card goes in drop box.

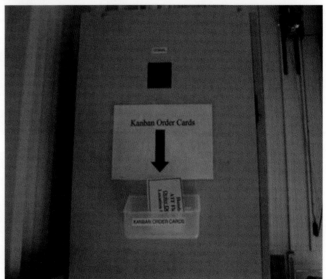

- Signal: "I have an order."
- Material handler reacts to signal.
- No signal? No need to react.

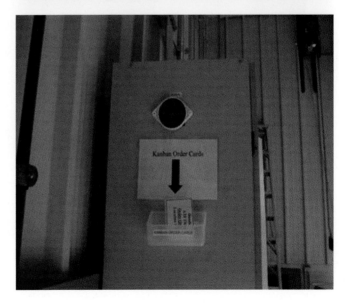

Supplies Are Delivered and Card Returned

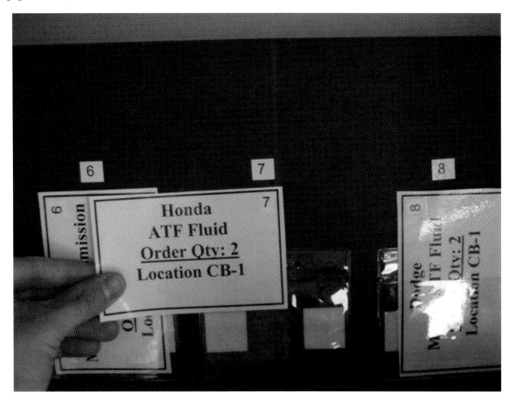

- Kanban cards must have a number.
- Cards can return to location.

- Box Kanban card

- White card is pulled for ordering.
- Red Card stays: Visual Indicator = Order Has Been Placed.

Maintenance Kanban Cards

- Cards contain pictures for quick reference.

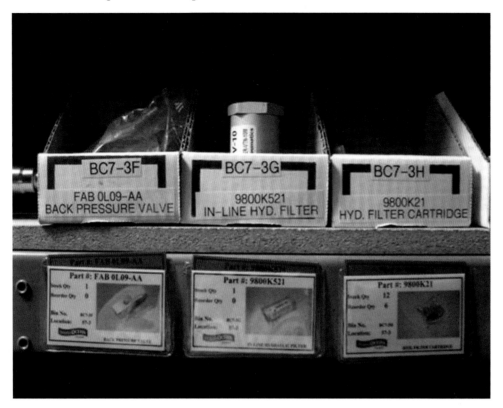

- Parts are organized based on machine type.

Chapter 4
Two-Bin Systems and Material Handling

Two-Bin Systems

Two-bin systems are an effective Kanban method for handling small parts and hardware. The great thing about a two-bin system is that you do not need to create Kanban cards. In a work area, each part using the two-bin system has two bins for storage. Each bin has the same amount of inventory. An operator simply works out of one bin at a time; when that bin becomes empty, the entire bin is placed in a replenish drop or area. The empty bin is the Kanban as the bins contain a label with all the required information that you would find on a Kanban card. Two full bins represent the maximum quantity desired in the work area; the second bin is the minimum, so when the bin is empty, the worker has reached the minimum. It is now time to order. The operator then continues to work out of the second bin as the first one is replenished. It is also good to install some form of a communication light that can be turned on to signal to a material handler that there is an empty bin to be picked up.

Another good rule of thumb is to color code the bins or labels of each individual work area. This provides a quick visual reference that either the correct or incorrect bin is in a location.

- Two-bin system is set.

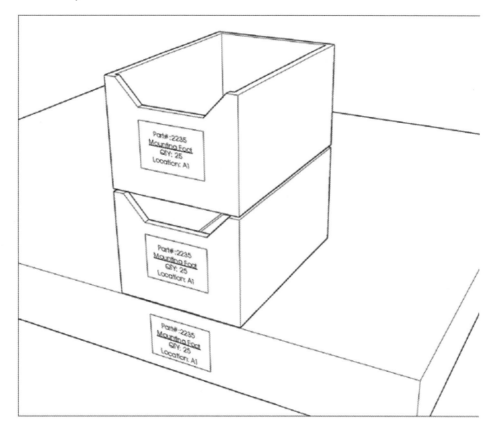

- Empty bin is removed and placed in replenishment drop.

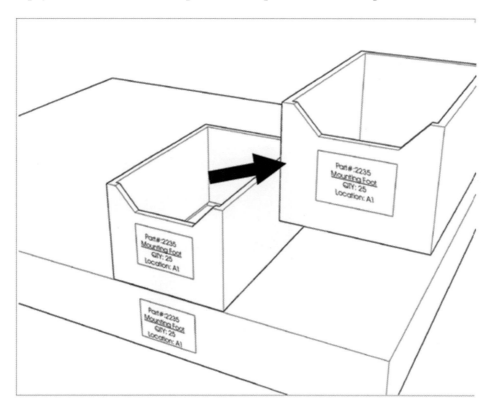

- Worker proceeds to work out of remaining bin.

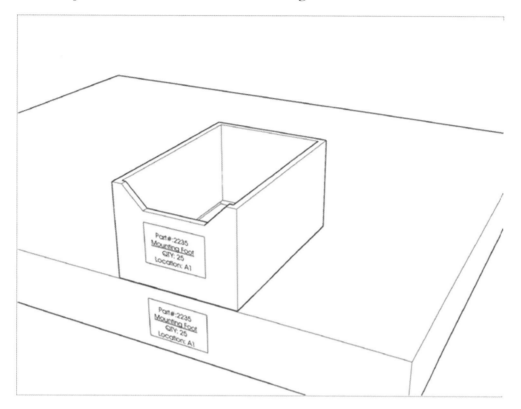

- Full bin returns and is placed under the first for first in, first out (FIFO).

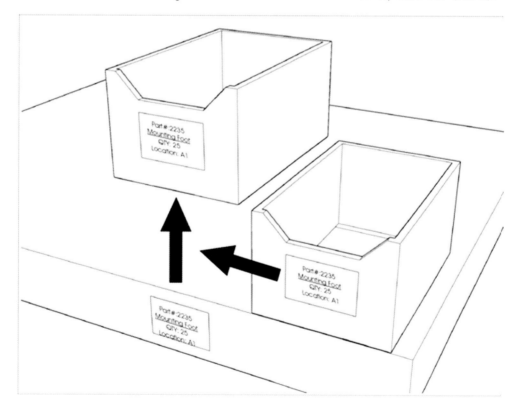

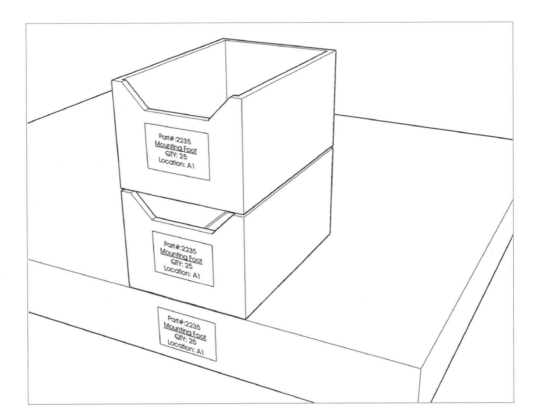

- Two-bin system continues to run.
- Operator never left the work area.
- There are *no* part shortages.
- Inventory is controlled through the pull system.

- Green labels are on bins.
- Label on the rack identifies the home location.
- All three labels are identical and match up.

- The second bin is behind the first.
- The material handler fills from the back.

Material Handling

Material handlers are a vital component of any Lean process. Generally deployed into production areas, material handlers keep material flowing in and out of the work area, allowing the workers to focus on performing value-added work. Many companies view material handlers as a waste of company resources as their work is considered indirect. Direct labor creates value for the company by producing product that can be sold and brings in money. As indirect labor, material handlers' contributions only take money from the company. Although there is truth to these statements as far as direct and indirect are concerned, allow me to paint you a different picture.

Imagine a race team. There is a driver and a pit crew. There are other people/employees of a race team, such as those in management, but let us stay with the driver and pit crew. The driver can be looked at as direct labor. His or her job is to race the car and complete the required number of laps in the least amount of time compared to the competition. The act of racing is value added. Imagine if the driver had to get out of his or her car and perform the actions at the pit stop: change tires, fill the gas tank, clean the window, and so on. Seems silly, does it not? The race is lost. The pit crew acts as the material handlers of the "system," and their contributions are valuable. Also, if you were to take three videos of three pit stops from one race and place them on a screen at the same time, their movements and time would be virtually identical. There is structure to what they do, where they go, the steps taken, and so on.

Returning to direct and indirect labor, many companies reduce or eliminate material handlers in an attempt to reduce indirect labor costs. These costs are only seen on accounting ledgers. Line workers (direct labor) are now required to retrieve their own material to conduct their work. Although this movement is not tracked, the second a direct labor worker leaves his or her work area, the worker has just become indirect labor.

If the movement is needed, why would you have operators perform the work? You are losing output, reducing productivity, increasing lead times, and decreasing capacity—and possibly creating potential quality problems because the operators are now redirecting their focus multiple times.

Incorporating material handlers is part of the game; more important is the system they work within. This section helps illustrate the material system using Kanban cards and the two-bin system to show you how it works.

Communication Lights

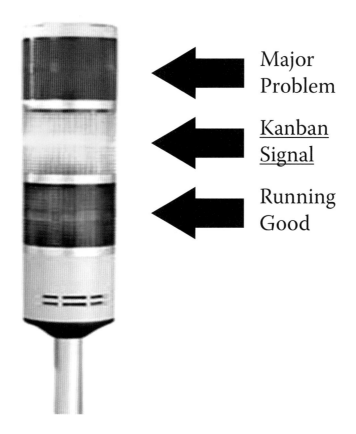

← Major Problem

← Kanban Signal

← Running Good

- Lights act as communication to everyone
- Great for communication to material handling
- Lights: Signals for communicating empty bins or cards in a workstation

Kanban Card or Bin Drop Point

- Operators place empty bin or card
- Turn on light to signal material handlers
- Material handlers come to drop point
- Retrieve cards and bins
- Turn light off
- Return with parts material

Drop points can be strategically placed through the work area, and each operator on the production line can be assigned to a drop point. By assigning operators to drop points, you keep the flow of material in and out even. This ensures you do not clutter drop points and do not overload material handlers.

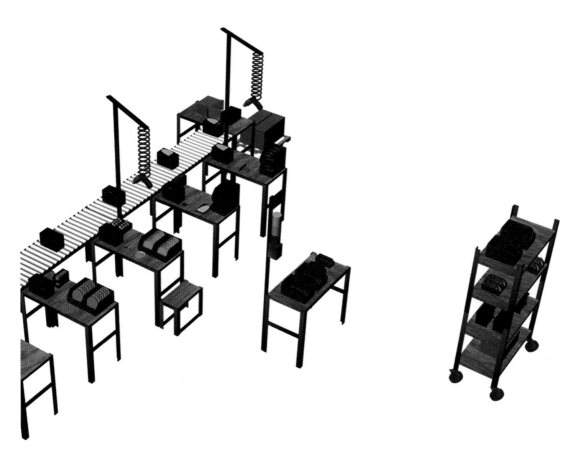

In many cases, the progress and flow of the line can be gauged by the flow of the material-handling system. Because the quantities of parts in the work area are based on a rotation with minimum and maximum quantities, parts should be flowing in and out based on output. If there are any issues with output, signals for replenishment will not appear as planned.

- Station location
- H3 displayed on bin label

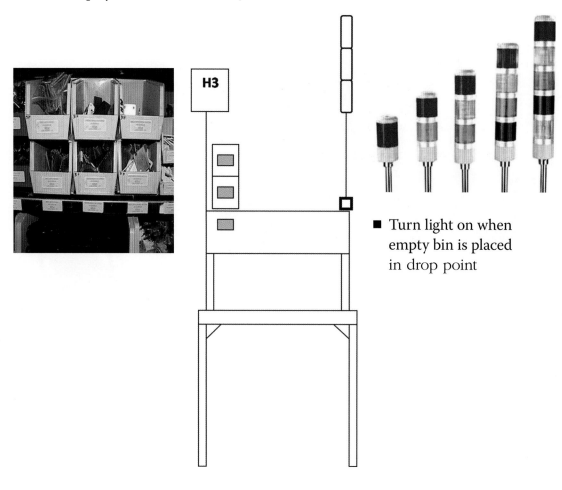

- Turn light on when empty bin is placed in drop point

Chapter 5

In-Process Kanban

What Is a Pull System?

- Production of items as demanded or needed by external and internal customers
- Production of items as needed by other processes

External and Internal Customers

- External customers: These are your actual customers purchasing your final product. External customers trigger work in the factory.
  - Dealers, distributors, end users
- Internal customers: These are manufacturing or support processes in the facility, sending parts, material, and supplies to each other. Internal customers then trigger work within each production process.
- These processes provide necessary items to each other so they can perform work:
  - Fabrication
  - Painting
  - Assembly
  - Welding
  - Packaging
  - Stockroom or warehouse
  - Production control

Process Based (Push)

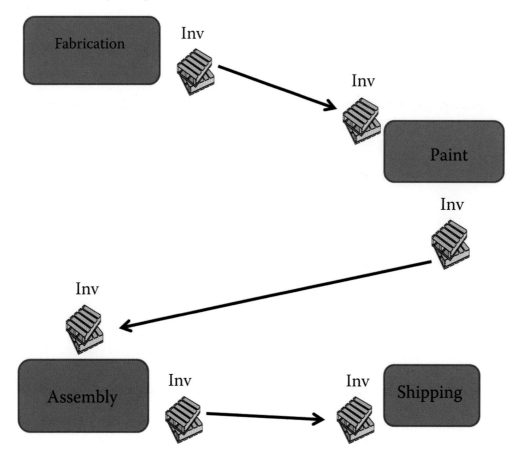

In-Process Kanban (Pull)

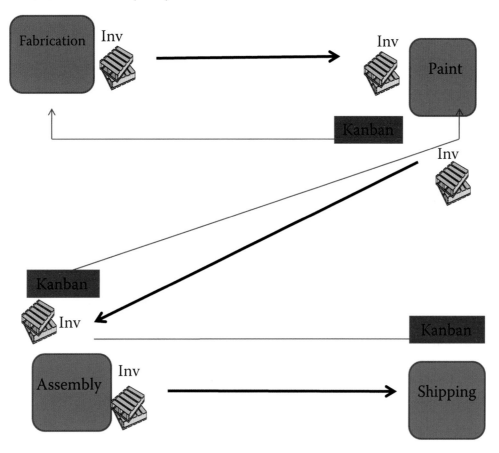

- Example: Production processes are now making product when "signaled" by a Kanban communicating the need for product

Pull System from Assembly to Paint

- Staged painted parts ready for assembly

Items Have Location on Rack

- Picture, PN (part number), Paint Qty (paint quantity), and Location

Information and Signal

- Model, station of consumption.
- Green: Rack is full and ready to be pulled.
- Location for staging and ID.

How It Works

- Green: Full and ready for assembly
- Assembler depletes rack of parts
- Turns sign to red (signal)
- Carts are switched
- Painter looks for red only
- Pulls rack
- Reads cards and paints
- Turns sign to green

In-Process Kanban ■ 43

- 454: Model number
- SA1: Station number

- Same designation on floor for proper return

- In-Process Kanban (IPK) states all required information

- Quick visual that the part was painted the wrong color

- Simple IPK staging
- Once the rack is full, stop!
- When part is pulled, start!

- IPK in outgoing saw area
- Next station pulls and welds
- Full = Stop!
- Empty Space = Start!

Welding into Powder Coating

- IPK carts

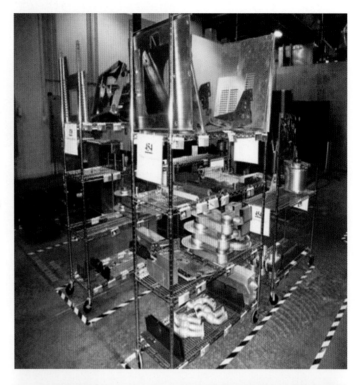

- Maximum: Three parts
- Green: Full
- Red: Replenish
- Reduces overproducing
- Work based on priority

IPK into Powder Coating Department

- FIFO (first in, first out) lanes
- Model specific
- Dollys
- Reduce cycle time
- Reduce strain

Writable In-Process Kanban (IPK)

Saw Operation

- Card is filled out by the consuming process needing parts.
- The IPK card is assigned to a saw.
- Saws have priority orders.
- Saw operator takes card.
- Cuts are based on the card information.
- Length, size, order number, part number, and so on

Conclusion

Successful Kanban implementations can have a profound impact on all the Lean metrics outlined in this playbook. Carrying excessive levels of inventory ties up money that could be used for purposes that are more beneficial to the company. Not every company is the same, and you must look at each part and supply individually to determine what the best replenishment system will be to implement better control. Start with shop supplies and then work your way to raw parts and material. Start small as well and maybe focus on a couple of supply cabinets first, then branching into an area with more inventory.

You will refine your Kanban systems every year by evaluating the changes in demand and usage. You may also find that your first implementation may have miscalculated quantities, and you will need to improve yet again. This is good, and I would not expect anything different. I recommend challenging your maximum and minimum quantities as needed to help create a flexible but robust Kanban system.

I hope this playbook has provided the hands-on detail you need and will allow you to get started. My final thought is that you do not forget the power of material handling and creating a mechanism for moving material and parts through the whole company. Material handling becomes the monitoring system for all the Kanban systems, and the time and money invested in resources for this will be returned in dramatic fashion.

Good luck!

Definition of Terms

Communication lights: Used for many purposes; however, in a material replenishment system, this light is used to signal to material handlers that an order has been placed for more parts for a work area.

Defects: Mistakes made in the process requiring rework and creating material scrap and lost products.

Floor space: Performance measurement of how much factory space is being used to conduct value-added work. Often measured in profit per square foot or revenue per square foot.

In-process Kanban: A visual reordering system used to trigger the need to move or build WIP (work in process) through a factory.

Inventory: The measurement in quantity and cost of raw material, WIP, and finished goods. Higher than needed inventory levels due to excessive purchasing of raw material, overproducing WIP, and unsold finished goods. Inventory ties up working capital, takes up floor space, and adds to longer lead times.

Kanban: Japanese work meaning "signal."

Kanban card: A visual reordering card placed near parts and material; the card contains information on how to order.

Motion: Movement of workers generally leaving their work areas to find items unavailable in the work areas.

Overprocessing: The act of overperforming work steps, such as redundant effort extra steps.

Overproduction: The act of producing more product than necessary, performing work in the wrong order, and creating unneeded inventory.

Productivity: One of the six Lean metrics that is a measurement of a worker's efficiency in a process. Often, it is a comparison of the time allocated to perform work to the actual time the worker took to perform it.

Quality: Internal measurement of rework, scrap, and defects in a production process.

Throughput time: Time associated with all value-added and non-value-added time in a process. It is the time it takes material to get through the first to the last steps of the entire factory. Raw material to finished goods.

Transportation: The movement of raw, WIP, and finished goods throughout the company.

Travel distance: Measurement of the physical distance product and worker go and the time associated with it. A long travel distance equates to longer lead times in the process.

Two-bin system: A visual Kanban system in which parts or material are placed in two bins. Once a bin is empty, it acts as a Kanban.

Waiting: When work comes to a stop due to lack of necessary tools, people, material, information, and parts. Wait time is often called queue time.

Wasted potential: Poor use of people, including skill sets not being utilized, wrong job placement, and workers consumed in wasteful steps.

Index

A

addresses, 13
annual evaluation, 49
assembly, pull *vs.* push system, 38–39
assembly to paint example
    information, 41
    items, location on rack, 40
    operation of system, 42–45
    signal, 41
    staged parts, 40

B

background information, *ix–xv*
bin drop point, material handling, 34–36
boardroom Lean, *ix*
body shop supplies, 11
bottle quantities, 18
buffers, 19
building cost, *xiv*

C

cabinets, 9, 20, *see also* Storage
calculation, new on-hand inventory costs, 5
carrying costs, *xiii*
categories
    Kanban sizing report, 2
    parts, 17
cell manufacturing, *x*
cleaning, overprocessing, *xi*
color coding
    assembly to paint example, 41–42, 46
    Kanban cards, 24
    red dots, 20, 22
    red tags and red tagging, 7
    two-bin systems, 27

communication lights
    defined, 51
    drop points, 36
    material handling, 33
    two-bin systems, 27
cons, *see* Pros and cons
costs
    excessive inventory, 49
    per unit, 2
    savings, 5
    total on hand, 2
cycle counting and baseline
    basic concepts, *x*, 1
    calculation, new on-hand costs, 5
    cycle counting, 3
    identification, new inventory quantities, 4
    Kanban sizing report, 2, 6
    on-hand inventory, 3
    part identification, 3
    vendors, 3
cycle time waste, *xii*

D

deburring, overprocessing, *xi*
defects
    defined, 51
    floor space waste, *xi*
    painting, quick visual, 44
    as waste, *xii*
delivery
    system for supplies, 23–24
    throughput time, *xv*
doors, lack of, 11
drop points
    communication light, 36
    Kanban cards, 22

material handling, 34–36
two-bin system, 28

E

employees, *see* Wasted human potential
enterprise resource planning (ERP) software, 6
ERP, *see* Enterprise resource planning software
errors, *see* Defects
evaluation, annual, 49
excessive amounts
    handling, *xiv*
    inventory, 9
    materials, 10
    supplies, 8
external customers, 37
external supply chain, *x*
extra steps, *see* Overprocessing

F

fabrication, pull *vs.* push system, 38–39
first in, first out (FIFO), 29, 47
5S methodology
    basic concepts, *x*, 7
    Kanban card system, 17
    location designations, 13
    sorting, 8–12
floor space
    defined, 51
    inventory, *xiii*
    Lean metrics, *xiv*
    location designation on, 43
    longer-than-needed processes, *xv*
    shadow boards, 12
    travel distance, *xv*
    waste, *xi*
flow, gauging, 35
foam tape example, 15

H

handling, excessive, *xiv*
hardware
    categories, 17
    storage, 12
    two-bin systems, 27
human potential, wasted, *xii*, 52

I

identical labels, 31, *see also* Labels
identification, new inventory quantities, 4

idleness, *see* Waiting
imbalanced workloads, *xii*
implementation
    basic concepts, *ix–x*
    matching Kanban cards, 3
    sorting, 8
    starting small, 49
information, assembly to paint example, 41
in-process Kanban (IPK)
    defined, 51
    external customers, 37
    in-process Kanban system, 39
    internal customers, 37
    powder coating department example, 47
    pull system, 37–45, 47
    push system, 38
    saw operation example, 48
    welding into powder coating example, 46
    writable in-process Kanban, 48
insurance costs, *xiii*
internal customers, 37
inventory, *see also* Materials
    calculation, new on-hand costs, 5
    defined, *x*, 51
    establishing new levels, 4
    excessive levels, 9, 49
    on hand, 3
    identification of quantities, 4
    Lean metrics, *xiii–xiv*
    pull system control, 30
    shadow boards, 12
    as waste, *xii*
IPK, *see* In-process Kanban

K

Kanban
    background information, *ix–xv*
    basic concepts, 15
    benefits, *xiii*, 15
    defined, 16, 51
    implementation, *ix–x*
    review, 49
    waste reduction, *xii*
Kanban card system
    basic concepts, *x*
    breaking into categories, 17
    defined, 51
    examples, 21, 24
    5S implementation, 17
    locations, 13
    maintenance cards, 25
    matching during implementation, 3

material handling, 34–36
maximum quantities, 18
minimum quantities, 19
number, 23
overview, 17
pictures for reference, 25
reorder signal/quantity, 19–20
right-size area, 17
supplies, system for, 21–24
usage, 18
Kanban sizing report
cycle counting and baseline, 2–3
location designations, 12
sorting, 11
key performance indicators (KPI), *xii*, *see also* Lean metrics

L

labels, two-bin systems, 27, 31, 36
lead times, *xv*
Lean metrics, *xii–xv*
lights, *see* Communication lights
locations
assembly to paint example, 40–41
designations, 13
drop points, 34
Kanban cards, 15, 23
stations, 36
two-bin systems, 31
longer production lines, *xv*

M

maintenance Kanban cards, 25
maintenance shop supplies report, 2, 4
maintenance-tracking software, 6
manufacturing building cost, *xiv*
material handling
basic concepts, *x–xi*, 32
bin drop point, 34–36
communication lights, 33
importance, 32, 49
Kanban cards, 34–36
material requirements planning (MRP) software, 6
materials, *see also* Inventory
categories, 17
excessive, 10
maximum quantities
5S methodology, 7–13
identification, new inventory quantities, 4
Kanban card system, 18

storage needs, 11
two-bin systems, 27
measures and metrics, *x, xii–xv*
Microsoft Excel software, 2
minimum quantities
identification, new inventory quantities, 4
Kanban card system, 19
two-bin systems, 27
mistakes, *see* Defects
model number, 43
motion and movement
defined, 51
material handlers, 32, 49
as waste, *xii*
MRP, *see* Material requirements planning software

O

obsolescence, *xiii, xiv,* 10
on-hand inventory, 2
operation of system, 42–45
Ortiz, Chris, 55
output issues, 35
overprocessing, *xi,* 51
overproduction
defined, 51
parts, 10
reduction, 46
as waste, *xi*

P

packaging station, 13
painting
assembly to paint example, 40–45
pull *vs.* push system, 38–39
parts
categories, 17
cycle counts, 3
description, 2
extra, from sorting, 10
identification, 3
number, 2, 40
organized by machine type, 25
shadow boards, 12
shortages, two-bin system, 30
tagging, 3
physical obstacles, *xiv*
pictures for reference, 25, 40
pit crew example, 32
planning, endless cycles, *ix*
point of use, *xii*

polishing, overprocessing, *xi*
priorities, work based on, 46
production lines, long, *xv*
productivity, *xiii*, 51
profit per square foot, *xiv*
progress, gauging, 35
pros and cons, quantities, 18
pull system
   assembly to paint example, 40–45
   basic concepts, 37
   controlling inventory, 30
   external customers, 37
   in-process Kanban system, 39
   internal customers, 37
   IPK into powder coating department example, 47
   saw operation example, 48
   welding into powder coating example, 46
   writable in-process Kanban, 48
push *vs.* pull system, 38–39

Q

quality
   defined, 51
   floor space waste, *xi*
   Lean metrics, *xiii*
   potential problems, lack of focus, 32

R

race team example, 32
racks and shelving
   addresses, 13
   adjustable shelves, 12
   floor space, *xiv*
   identification, 13
   inventory, *xiii*
reassembly area, 21
rechecking/reverifying, *see* Overprocessing
red dots, 20, 22
red tags and red tagging, 7–8, *see also* Color coding
reorder signal/quantity, *see also* Kanban; Signals
   identification of new quantities, 4
   Kanban card system, 19–20
   red dots, 20
replenishment drop, 28
returning cards, 23–24
revenue, creating, *xiv*
review, 49
reviews, annual, 49
rework, *see* Defects

right-sizing
   area, 17
   cabinets, 20
   phase, 8

S

sanding, overprocessing, *xi*
savings, *see* Costs
saw operation example, 45, 48
scrap, *see* Defects
scrub, *see* 5S
set in order, *see* 5S
shadow boards, 12
shipping, pull *vs.* push system, 38–39
shortages, two-bin system, 30
signals, *see also* Communication lights; Kanban; Reorder signal/quantity
   assembly to paint example, 41
   red dots, 20, 22
small parts, two-bin systems, 27
sorting, *see* 5S
space, poor use of, *xiv, see also* Floor space
staging, assembly to paint example, 40–41, 45
standardize, *see* 5S
stations
   of consumption, 41
   locations, 36
   number, 43
storage, 11–12, *see also* Cabinets
subassemblies, obsolete, 10
supplies
   basic concepts, 21–22
   body shop, 11
   categories, 17
   cycle counts, 3
   delivery, 23–24
   excessive, 8
   productivity, *xiii*
   returning card, 23–24
supply chain, external, *x*
sustain, *see* 5S

T

tagging parts, 3
throughput time, *xv*, 51
total cost on hand, 2
training, endless cycles, *ix*
transportation, *xii*, 52
travel distance
   defined, 52

Lean metrics, *xv*
productivity, *xiii*
two-bin systems
basic concepts, *x*, 27–31
defined, 52

U

usage, 18
"use-up" pile, 8–9
utilizing employees, *see* Wasted human potential

V

value-added work, *xiv*, 32
vendors, 2–3
visual storage, 11

W

waiting
defined, 52
travel distance, *xv*
as waste, *xii*
walking, *see* Travel distance
warranty claims, *see* Defects
wasted human potential, *xii*, 52
wastes, *x*, *xi–xii*
welded parts, extra, 10
welding into powder coating example, 46
white Kanban cards, 24
work area design, *xv*
workload imbalance, *xii*
workstations, *xv*
writable in-process Kanban, 48

About the Author

Chris Ortiz is the founder and president of Kaizen Assembly, a Lean manufacturing training and implementation firm in Bellingham, Washington. Chris has been featured on *CNN Headline News* on the show "Inside Business with Fred Thompson." He is the author of seven books on Lean manufacturing (see list that follows).

Chris Ortiz is a frequent presenter and keynote speaker at conferences around North America. He has also been interviewed on KGMI radio and the *American Innovator* and has written numerous articles on Lean manufacturing and business improvement for various regional and national publications.

Kaizen Assembly's clients include industry leaders in aerospace, composites, processing industries, automotive industries, rope manufacturing, restoration equipment, food processing, and fish processing.

Chris Ortiz is considered to be an expert in the field in Lean manufacturing implementation and has over 15 years experience in his field of expertise.

He is also the author of the following:

Kaizen Assembly: Designing, Constructing, and Managing a Lean Assembly Line (Boca Raton, FL: Taylor and Francis, 2006) (now in its second printing)

Lesson from a Lean Consultant: Avoiding Lean Implementation Failure on the Shop Floor (Upper Saddle River, NJ: Prentice Hall, 2008)

Kaizen and Kaizen Event Implementation (Upper Saddle River, NJ: Prentice Hall, 2009) (translated into Portuguese)

Lean Auto Body (Bellingham, WA: Kaizen Assembly, 2009)

Visual Controls: Applying Visual Management to the Factory (Boca Raton, FL: Taylor and Francis/Productivity Press, December 15, 2010)

The Psychology of Lean Improvements: Why Organizations Must Overcome Resistance and Change Culture (Boca Raton, FL: CRC Press and Productivity Press, April 2012): Winner of the Shingo Prize for Operational Excellence in Research, 2013

The 5S Playbook: A Step-by-Step Guideline for the Lean Practitioner (Boca Raton, FL: CRC Press, October, 2015)